Concept: Magnets attract some objects but not others.

Materials:	magnets	erasers	metal keys
	paper clips	hair pins	feathers
	nails	crayons	rubber bands
	corks		

Note: If your students are very young or immature, choose objects that are too large to swallow, or do the experiment under close supervision in small groups.

A.   Discussion:

Show all the objects to be used in the experiment.
"Do you think a magnet will pick up this_____?" "Why?"
You may want to record their predictions to refer back to when the experiment is over.

B.   Experiment:

1.   Students use their magnets to try to pick up each object.

2.   Record the results on the worksheet. Circle each object the magnet picked up. Cross out each object the magnet did not pick up.

C.   Share the results:

Let the children tell which objects they were able to pick up with their magnets. Put all the objects that were picked up together on a table. "How are all these the same?" "What did we find out about magnets by doing this experiment?" (Magnets pick up metal objects.)

Additional Activity:

Let children take the magnets around the classroom and schoolyard to find other objects that are attracted to a magnet.

Magnets pick up metal objects.

_____'s Experiment

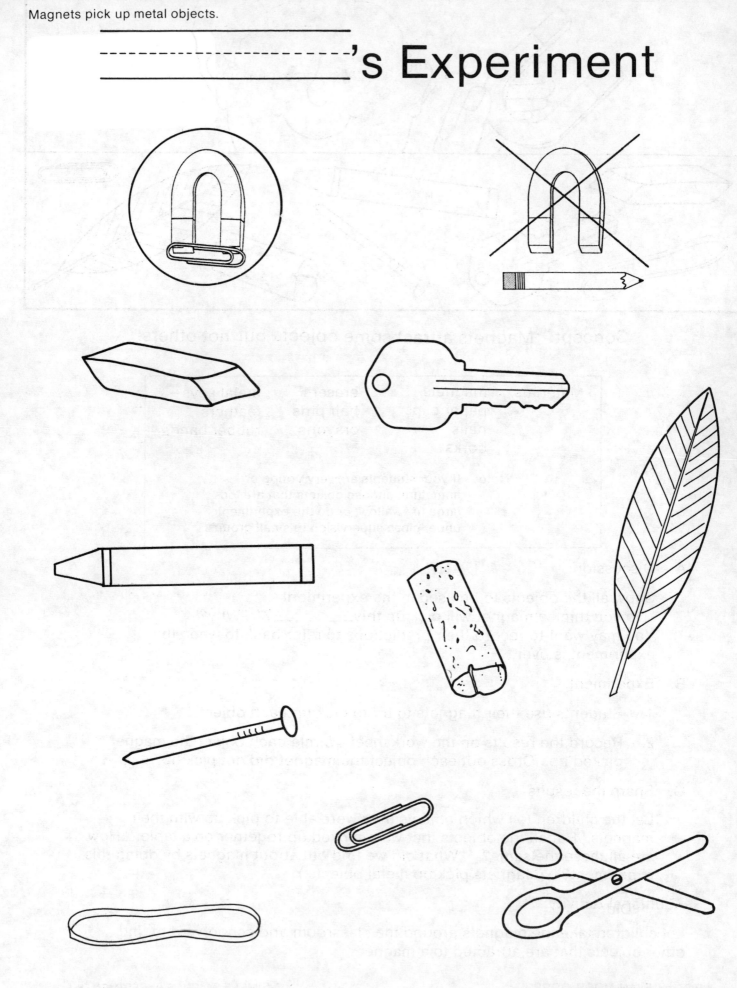

Concept: Magnets will attract some types of metal but not others.

> Materials: magnets   metal bottle caps
>   paper clips   metal spoons
>   hair pins   metal thimbles
>   nickels   brass screws
>   pennies   foil paper
>   nails
>
> Note: If your students are very young or
>   immature, choose objects that are too
>   large to swallow, or do the experiment
>   under close supervision in small groups.

A. Discussion:

Have children name and describe each of the objects.
"What are all of these made from?" (metal)
"Will your magnet pick up this_____?"
"Will a magnet pick up anything that is metal?"
You may want to record their predictions to refer back to after the
experiment is over.

B. Experiment:

1. Students try to pick up each object with their magnets.

2. Record the results on the worksheet. Circle objects that can be picked
   up. Cross out objects that cannot be picked up.

C. Share the results:

"Did your magnet pick up all the metal objects?"
Explain that only metals such as iron and steel are attracted to a
magnet.

A magnet will pick up objects made of iron and steel.

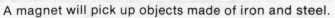

_____'s Experiment

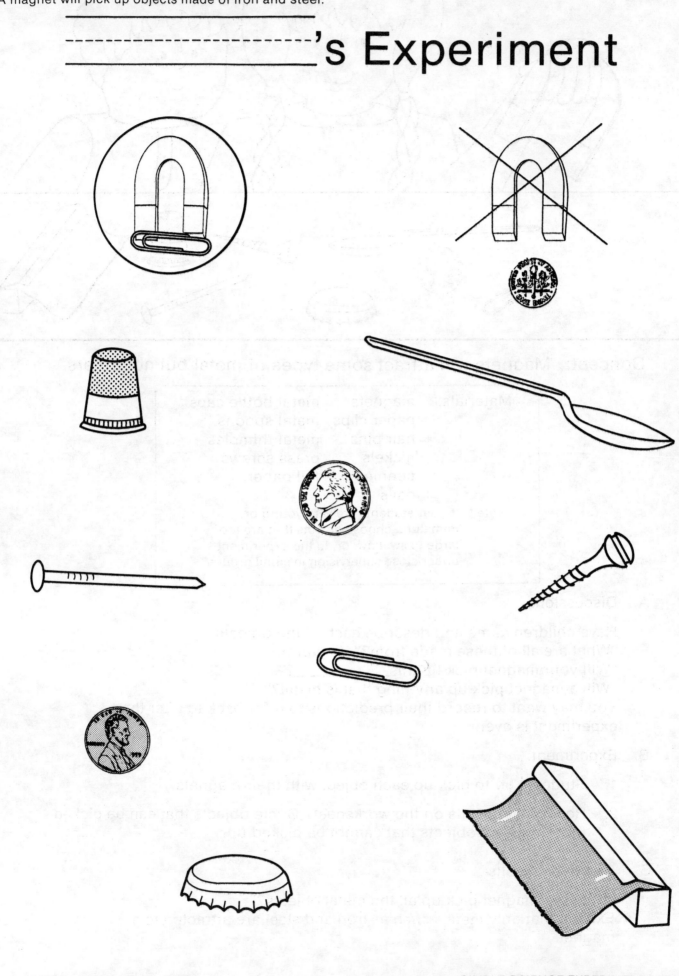

Concept: Some magnets are stronger than others.

Materials: magnets
paper clips
worksheets

A. Discussion:
Show all the different types of magnets you have available.
"Which magnet do you think will pick up the most paper clips? Why?"
"How many paper clips do you think this magnet (show one magnet at a time)
will pick up?"
You may want to record their predictions to refer back to after the experiment.

B. Experiment:
1. The student uses his/her magnet to pick up as many paper clips as possible.
2. Record the results on the worksheet. Count the number of paper clips. Cut
out the same number from the worksheet and paste to the picture of the
magnet.

C. Share the results:
Let each child tell (or show) how many paper clips were picked up.
"Who has the strongest magnet?"
"How many paper clips did the strongest magnet pick up?"
"Did we choose the correct magnet as the strongest?"

Some magnets are stronger than others.

_____
- 's Experiment
_____

My magnet picked up ☐ 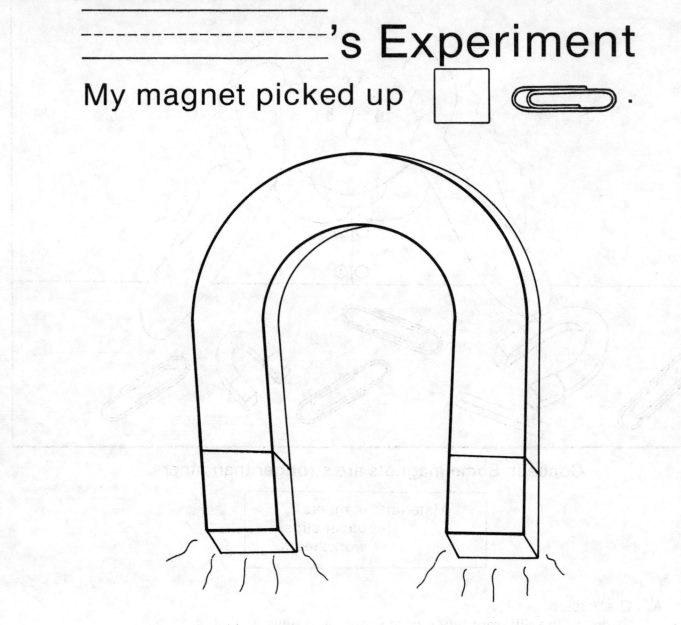 .

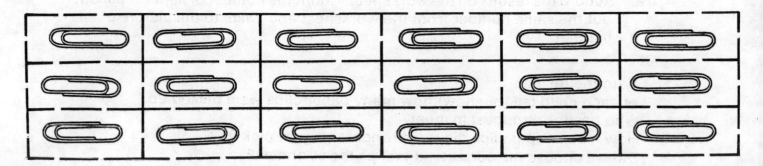

Concept: A magnet can attract metal through water.

Materials: magnets
paper clips
clear plastic cups
water

A. Discussion:

"Do you think a magnet can pick up a metal object in water? Why?"
You may want to record their predictions to refer back to after the
experiment.

B. Experiment:

1. Put water and paper clips in the clear plastic cup.

2. Stick the magnet into the water and try to pick up the paper clips.
   Record the results on the worksheet.

3. Put the magnet against the outside of the cup. See if your magnet will
   pick up the paper clips through the water *and* the plastic cup. Record
   the results on the worksheet.

C. Share the results:

Let the children tell what happened in the experiment.

Additional Activity:

"Will a magnet work in other liquids?" Try milk, juice, and syrup.

A magnet can attract metal through water.

_____

------------------------'s Experiment

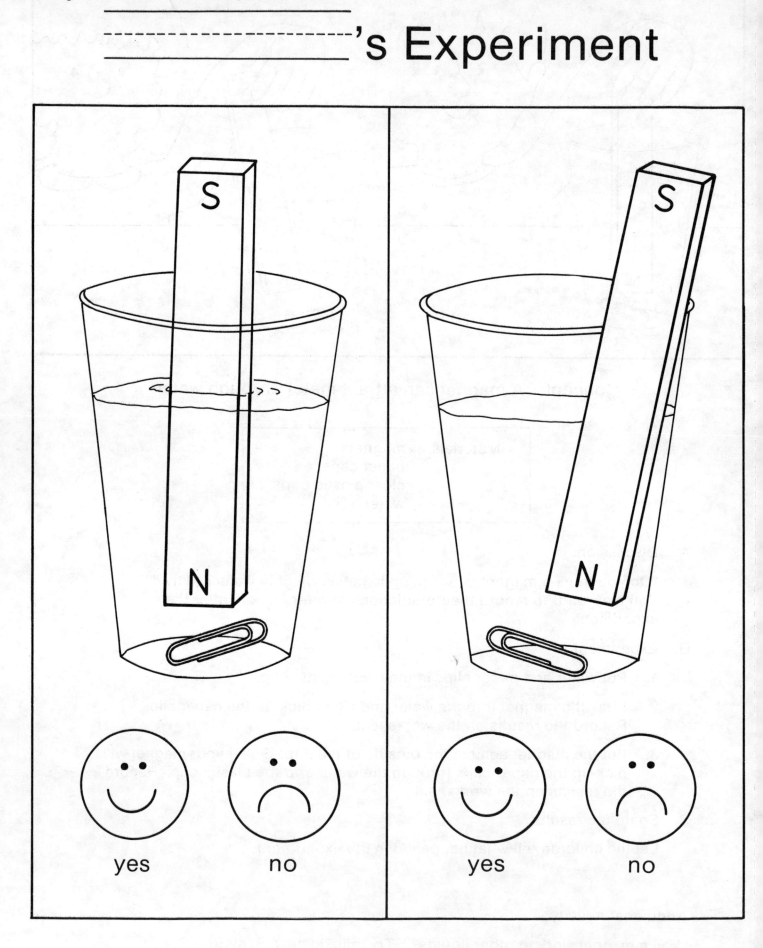

©1987 by EVAN-MOOR CORP. 8

Concept: A magnet can attract metal through other objects.

Materials:	magnets	drinking glasses
	paper clips	sheets of paper
	tongue depressors	scraps of cotton cloth

A. Discussion:

"Do you think your magnet will be able to move a paper clip that is
_____?" "Why?"

    (in a glass, under a sheet of paper, under a piece of cotton cloth, under a wooden stick)

You may want to record their predictions to refer back to after the experiment is over.

B. Experiment:

1. Children try each of the following experiments with their magnets:

    a. Put a paper clip in a glass. Move the magnet along the outside of the glass. Does the paper clip move?

    b. Put the paper clip under the sheet of paper. Place the magnet on the paper. Can you lift up the paper and the paper clip?

    c. Put the paper clip under the piece of cloth. Place the magnet on top of the cloth. Can you lift up the cloth and the paper clip?

    d. Put the paper clip on top of your piece of wood. Hold the wood and put the magnet under it. Move the magnet to see if you can get the paper clip to move.

2. Record the results on the worksheet by crossing out the smiling face if the magnet attracts the paper clip or the frowning face if it does not.

C. Share the results:

"Did your magnet move the paper clip through the_____?"

                         (glass, paper, cloth, wood)

Additional Activities:

Try the same experiment using thicker wood, cardboard, and heavier cloth.

A magnet can attract metal through some other objects.

_____

-----------------------'s Experiment

's Experiment

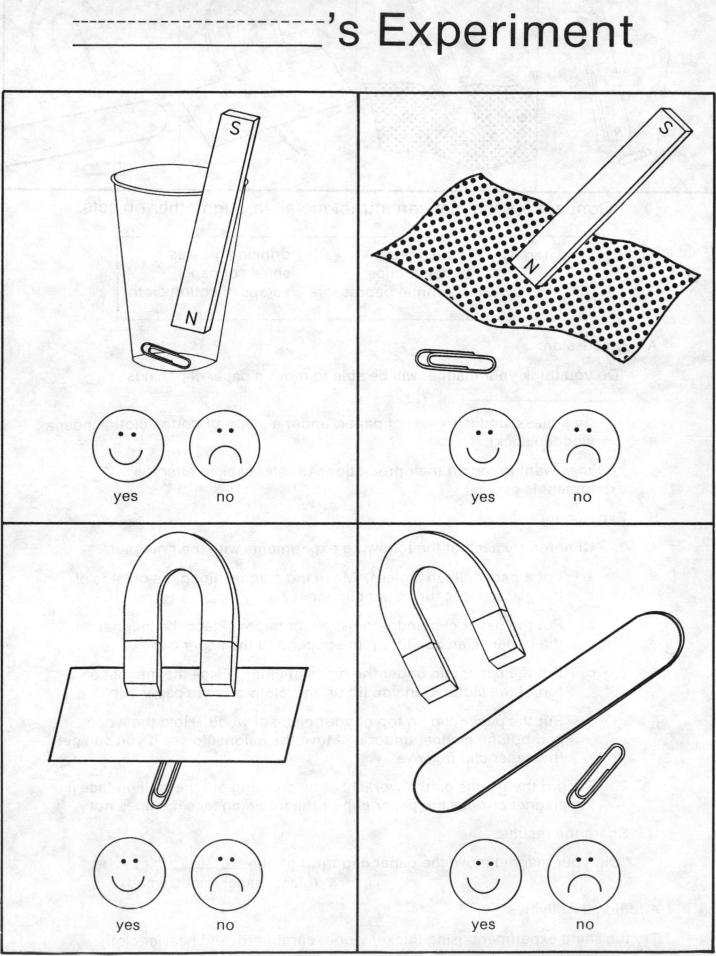

yes   no

yes   no

yes   no

yes   no

©1987 by EVAN-MOOR CORP.   10   SIMPLE SCIENCE EXPERIMENTS

Concept: A liquid takes on the shape of the container it is in.

> Materials: water
> containers of assorted shapes and sizes
> plastic tubs (to catch any water spills)

A.   Discussion:

"Water is a liquid. Does a liquid have a special shape? What shape is water? What happens to water when it is put into something? Do you think that this water will change shape if I put it into this *(name the container)* ? Let's find out."

B.   Experiment:

1.   Let each child or group of children pour water into several different types of containers to see what will happen.

2.   Record the results on the worksheet. Have children color water in each of the different shaped containers on the page.

C.   Share the results:

"Did the water change shape to fit into each container? Can you name other liquids that have the same shape as the bottle they are in?" (syrup, juice, milk, etc.)

Additional Activity:

Place an ice cube in a glass. "Is the ice cube the same shape as the glass?"

Let an ice cube melt. "Is the water from the ice cube the same shape as the glass?"

"Why is the melted ice cube the same shape as the glass, but the frozen ice cube is not?"

A liquid takes on the shape of the container it is in.

_____

--------------------------------'s Experiment

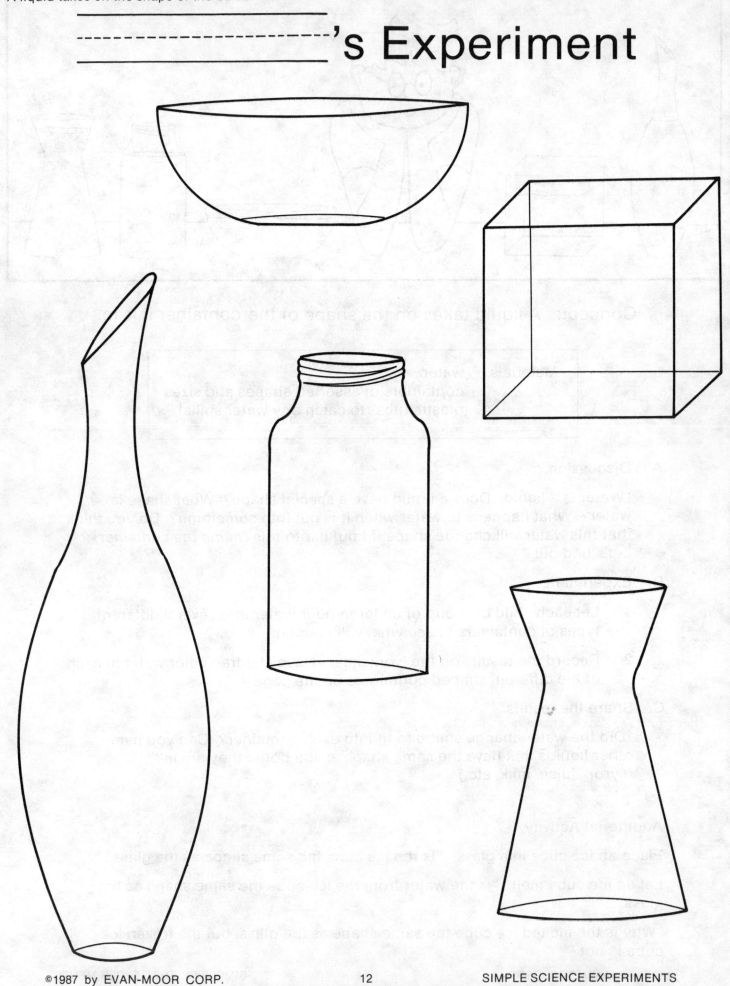

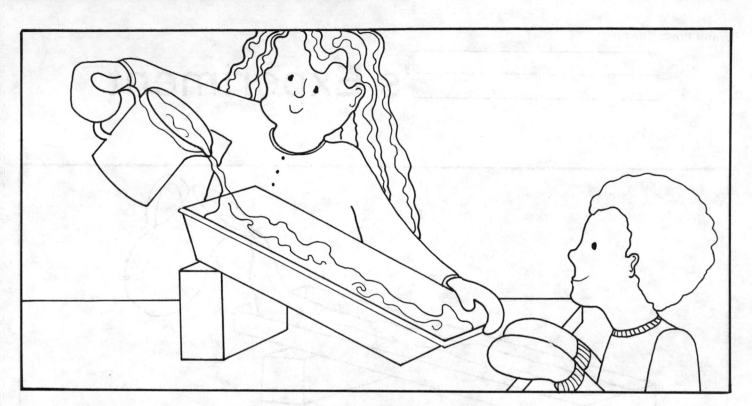

Concept: Water flows downhill.

> Materials:　containers of water
> plastic wash tubs (rectangular)
> large wooden blocks
> sponges and paper towels (for spills)
> blue crayons

A.　Discussion:

"Have you ever seen moving water? Where? Which way was the water going? Can you get water to move uphill? Will it stay there? Let's see what happens when we pour our water."

B.　Experiment:

1.　Have children work in small groups. Use wooden blocks to raise one end of the wash tub. Pour water into the tub at the high end. Observe how it flows.

2.　Have children try to figure out how to get the water to move uphill.

3.　On the worksheet show the movement of water by tracing the flow with blue crayon.

C.　Share the results:

"Which way did your water move? Did anyone find a way to get their water to move uphill? Did the water stay up? Why do you think water always moves downward?"

Water flows downhill.

_____

-----------------------------------'s Experiment

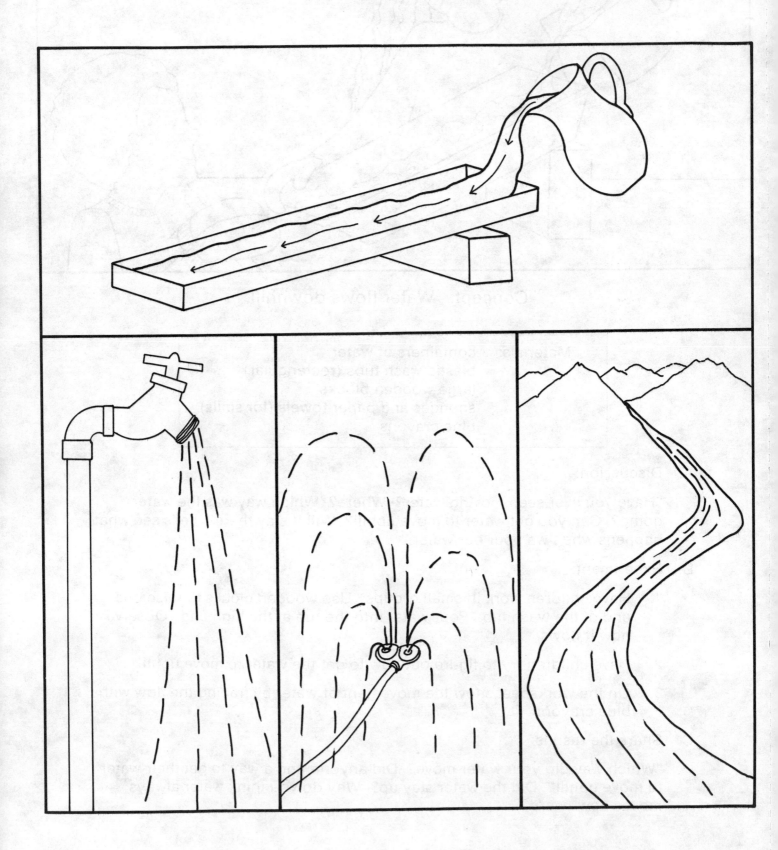

Concept: Moving water can do work.

Materials:	tag board or paper plates (6") pencils water plastic wash tubs (or sink)

A. Discussion:

"Can you tell me some of the ways you use the water? Do you think water can do work for people? How? We are going to do an experiment to find out one way water can do work."

B. Experiment:

1. Make a water wheel using the pattern on page 16.*
    Children cut the lines and fold each section over to form a "cup" to catch the water. Push the wheel onto a pencil. Wiggle the wheel until it moves freely around the pencil.

2. Hold the wheel under moving water from a faucet or container. (Be sure the folded "cup" sections are pointing up.)
    Watch the water turn the wheel.

C. Share the results:

"Did your water wheel turn? What made the wheel move?" Explain to your students that moving water can be used to make electricity.

*The teacher will need to trace the pattern on tagboard or use 6" paper plates. Then punch the holes and mark the cutting lines.

Moving water can do work.

_____

- - - - - - - - - - - - - -'s Experiment

_____

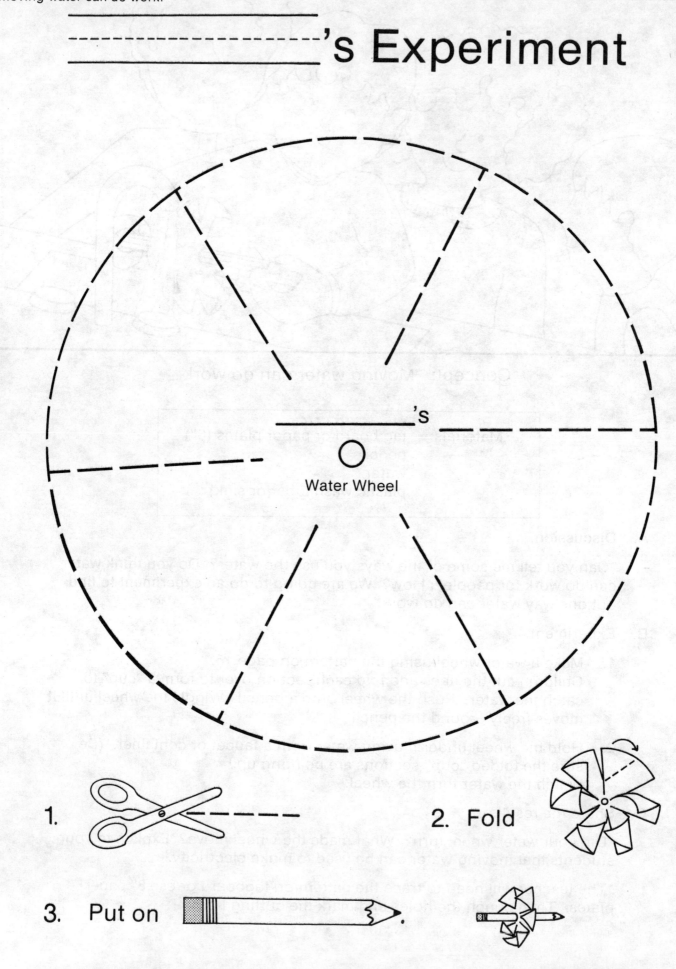

_____'s _ _ _ _ _ _

○

Water Wheel

1. ✂ - - - - - - - - -

2. Fold

3. Put on ▭▭✏.

©1987 by EVAN-MOOR CORP.   16   SIMPLE SCIENCE EXPERIMENTS

Concepts: We can change the form of a liquid by adding heat or cold.

Materials: access to a hotplate and freezer
              tea kettle
              ice tray
              water
              measuring cup

A. Discussion:

"Water is a liquid. Do you know a way we can change it into something else? What do you think will happen to it if we make the water very hot? What do you think will happen if we make it very cold?" You may want to record their predictions to refer back to when the experiment is over.

B. Experiment:

1. Measure the water as you put it into the tea kettle. Heat the water to boiling. Have children observe what happens as the water begins to boil. Cool the water and measure what is left. "What happened to the rest of the water?" (Be sure you let the water boil for several minutes to be sure there is a measurable difference.)

2. Put water in an ice tray. Put it in the freezer for several hours. Examine the ice tray to see what has happened to the water. "What change do you see in the water?"

C. Share the results:

"What happened to the water when we made it hot? Do you know what we call the water when it goes into the air? (gas) What happened to the water in the freezer? What do we call the ice? (solid)"

Record the results of the experiment on the worksheet.

We can change the form of a liquid by adding heat or cold.

_____

----------------------------'s Experiment
_____

1. Draw the steam.

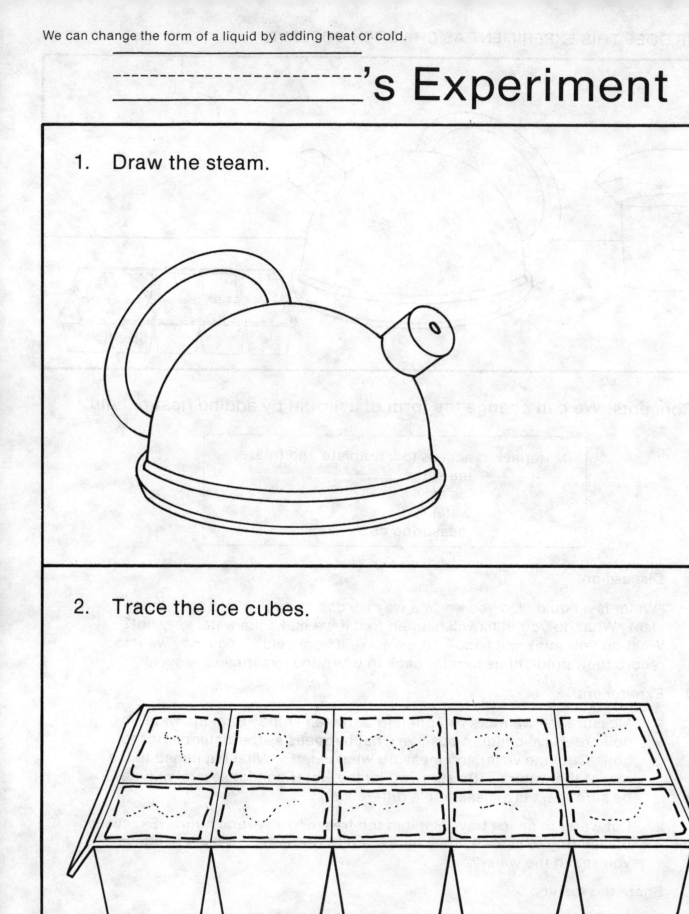

2. Trace the ice cubes.

©1987 by EVAN-MOOR CORP.   18

Concept: When water freezes it takes up more space.

Materials: jars or tall plastic glasses
masking tape
water
access to a freezer

A.   Discussion:

"We saw water turn to steam when we made it very hot and water turn to ice when we made it very cold. What did the ice look like? We are going to do another experiment to find out about water and ice."

B.   Experiment:

1.   Each child fills a glass or jar about half full of water. Mark the level with a piece of masking tape. (Write the child's name on the tape to identify his/her glass.) The child then colors water to the same level on glass #1 on his/her worksheet.

2.   Put the glasses in the freezer overnight.

3.   Return the glasses and worksheets to your students. They mark the level of the ice on glass #2 on the worksheet.

C.   Share the results:

"What happened to your water when it froze? Did the water stay the same size, get smaller, or get bigger when it turned to ice?"

Additional Activity:

Freeze other liquids to see if they also take up more space when frozen. (milk, juice, soup)

When water freezes it takes up more space.

_____

- 's Experiment

1

2

Concept: Ice melts.
(A solid can change into a liquid.)

Materials: plastic plates
ice cubes

A. Discussion:

"Is ice a solid or a liquid? Name some other solid things. Name some other
liquid things. What do you think will happen to our ice cubes if we leave
them out of the freezer? Let's see what will happen."

B. Experiment:

1.   Put an ice cube on your plate. Sit it in a warm place. Watch it to see
what happens to the ice.

2.   Record what happens by doing the cut and paste activity on the
worksheet.

C. Share the results:

"What happened to your ice cube? Why do you think it melted? Is it a solid
anymore?"

Additional Activity:

"Do all solids melt? Let's watch this popsicle, pebble, and crayon. Which ones
do you think will melt? Why? What other solids can you think of that will melt?"
(Example: butter, chocolate, ice cream)

Ice melts. (A solid can become a liquid.)

_____

- 's Experiment

_____

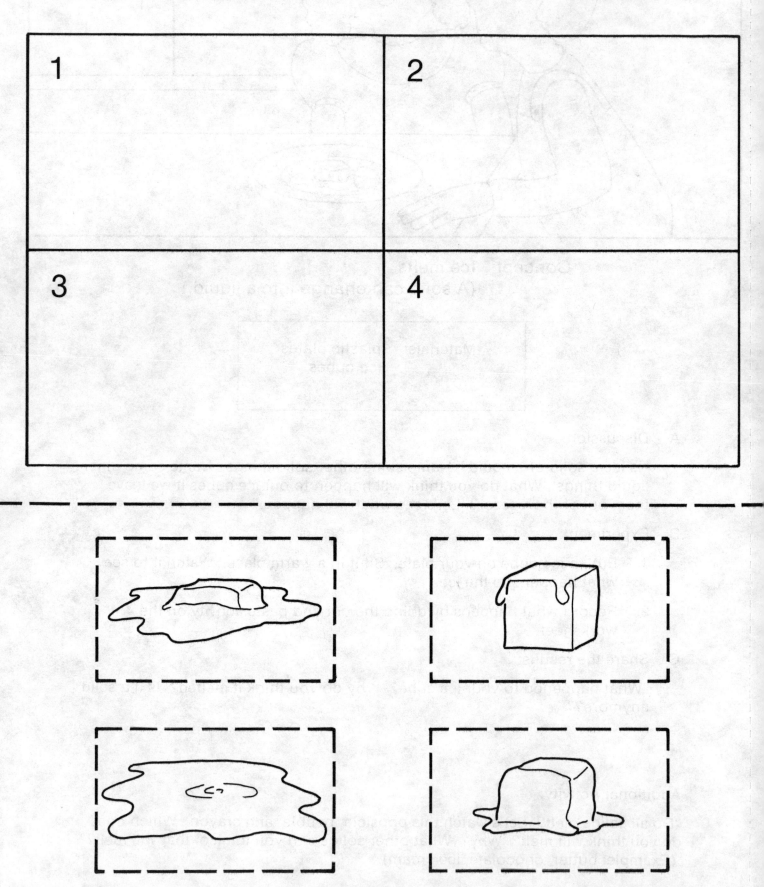

1	2
3	4

Concept: Small pieces of ice will melt faster than large pieces.
(There is more surface area exposed to heat.)

Materials: large ice cubes
small pieces of ice
plastic glasses
water

A. Discussion:

"We have learned that ice will melt. Will all ice melt the same? Do you think a big hunk of ice will melt faster or slower than little pieces? Why? Let's find out which one really melts faster."

B. Experiment:

1. Put water in two glasses. Put a large ice cube in one glass and several small pieces of ice in the other. Watch to see what happens.

2. Record the results on the worksheet. Circle the glass that shows which ice melted faster.

C. Share the results:

"Did the little pieces of ice melt faster or slower than the large ice cube?"

Additional Activity:

"Do you think the water helped make the little pieces melt faster? Let's put some little pieces in water and some more little pieces on a dry plate and see which one will melt faster."

©1987 by EVAN-MOOR CORP.   SIMPLE SCIENCE EXPERIMENTS

_____'s Experiment

Which melts faster?

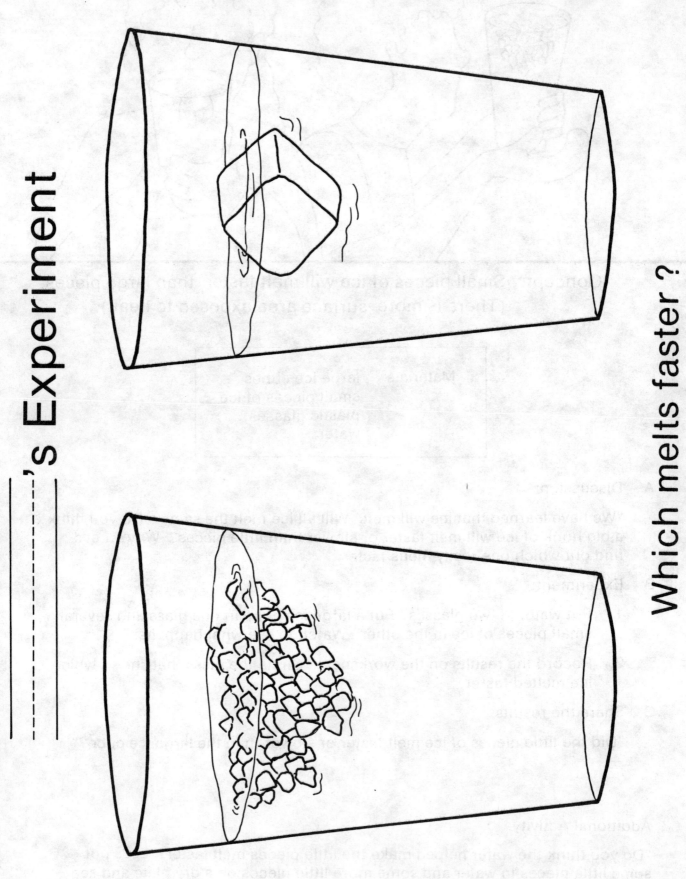

Small pieces of ice melt faster than large pieces. (More surface area is exposed.)

Concept: There is water in the air.

Materials: jars with lids
            ice cubes
            paper towels

Note: If you cannot find enough jars with lids,
       use plastic cups covered with plastic wrap
       held in place with rubber bands.

A. Discussion:

"There is water in the air. Can you think of a way to get the water out of the air so we can see it? Let's try this experiment."

B. Experiment:

1. Put several ice cubes in the jar. Put the lid on the jar and wipe off the outside with a paper towel.

2. Let the jar sit in a warm place for a little while. Look at the jar to see what has happened.

3. Record what you see on the worksheet. Circle the jar that shows what happened.

C. Share the results:

"What did you find on the outside of your jar? Where did the water come from? Can you think of another place where you have seen water that came from the air?"

You may want to explain to your students that when the warm air touches the cold jar, the water in the air forms into drops they can see. (condensation)

There is water in the air around us.

_____

------------------------------'s Experiment

Concept: Water evaporates.

Materials: Clear plastic glasses
permanent marking pens
water
worksheet (1 for each student)
blue crayons

A. Discussion:
"What will happen to the water if we leave it sitting out in the room for several days?" You may want to record their ideas to refer back to after the experiment is completed.

B. Experiment:
1. Each child fills a glass about half full of water. He/she then marks the water level on the glass with the permanent marker and colors the same level on glass #1 on the worksheet using a blue crayon. Write the child's name on the glass.
2. Set the glasses in a warm place and wait 3 or 4 days.
3. Return the glasses and the worksheets to your students. They mark the new water level on the glass and color the same water level on glass #2 on the worksheet.

C. Share the results:
"What happened to the water in your glass?"
"Where do you think the water went?"
Give as detailed an explanation of evaporation as is appropriate for the age and ability of your students.
"Can you think of other places water could evaporate?"

Example: river
ocean
swimming pool
fish bowl

Children mark the picture at the bottom of their worksheet that answers the question.

Water evaporates.

_____

- - - - - - - - - - - - - - - - - - - 's Experiment

1   2

What happened to the water?

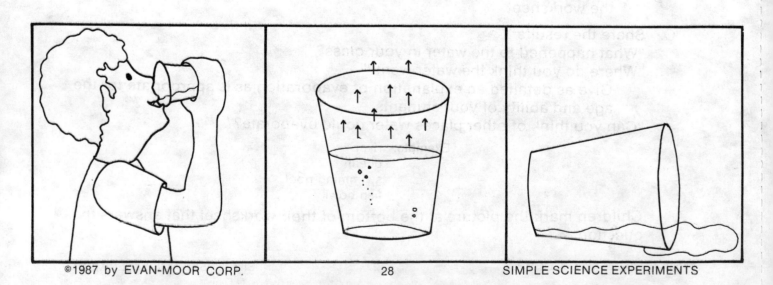

Concept: Some objects float and some sink in water.

Materials: plastic tubs   worksheets
          water   paper clips
          corks   plastic spoons
          marbles   rocks
          bits of wood   golf balls
          pencils

Note - If your students are very young or immature,
         use objects that are too large to swallow.

A. Discussion:
Show all the objects to be used in the experiment. "Do you think this____will
sink or float? Why?" You may want to record their predictions to refer back to
after the experiment is over.

B. Experiment:
1. The student puts each object in the tub of water to see if it will sink or float.
2. Record the results on the worksheet. Cross out the objects that sink. Circle
    the objects that float.

C. Share the results:
Let children share their discoveries.
"Why do you think a____sinks in water?"
"Why do you think a____floats in water?"
"Can you think of something else that probably floats/sinks in water?"

Some objects float and some sink in water.

_____

- - - - - - - - - - - - - - - -'s Experiment

_____

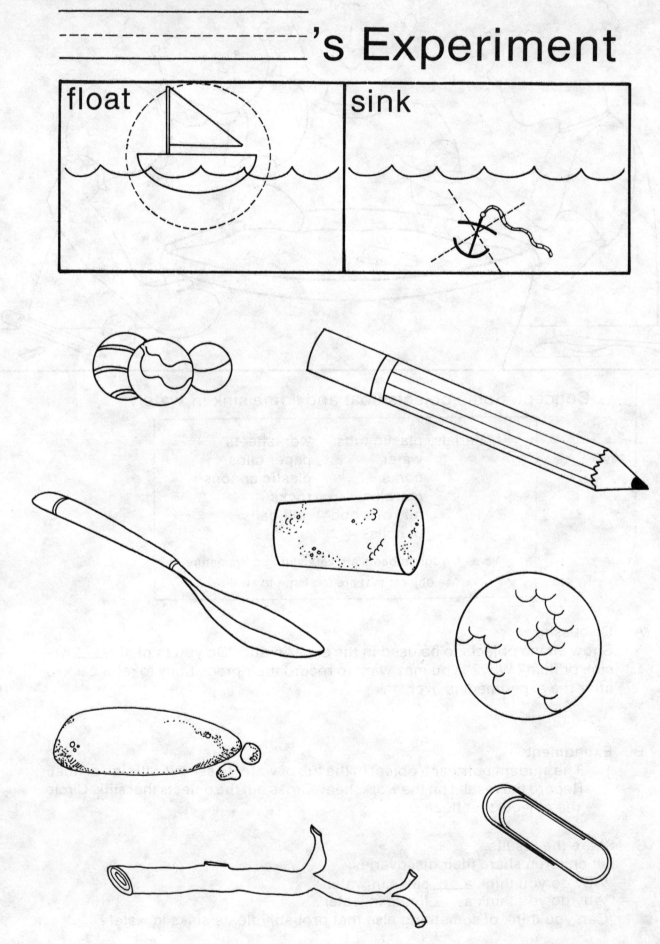

float

sink

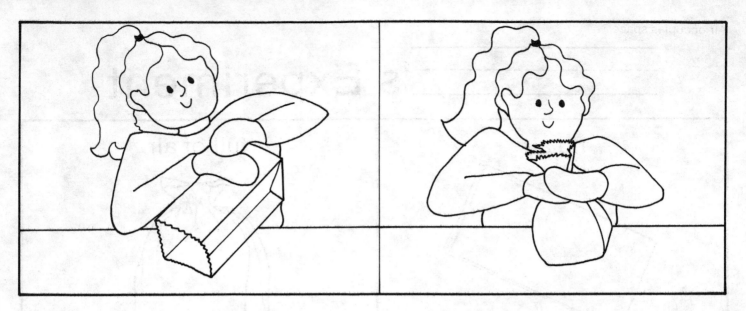

Concept: Air occupies space.

> Materials:　Children - paper lunch bags
> 　　　　　　Teacher - bicycle tire and pump
> 　　　　　　　　　　　 deflated beach ball
> 　　　　　　　　　　　 balloon

A.　Discussion:

"Where is air? Can you see it? Can you feel it? How can we tell that air is around us? In our experiment today we are going to find out more about air."

B.　Experiment:

1.　Children feel their lunch bags and look inside. "Is anything in your bag? Hold it open and blow into it. Hold the top tight with your hands. What is in your bag now?"

2.　Teacher demonstrates:
Put air into the bicycle tire, beach ball, and balloon. "We cannot see air around us, but we can see that air takes up space when we put it inside something."

3.　"Put your hands on your chest. Breathe in a big breath. Now let it out. Did you feel your chest go in and out? The air you breathe in fills your lungs up in the same way we filled the lunch bags, tire, balloon and beach ball."

4.　Do the cut and paste worksheet on page 32.

Additional Activity:

Go outside to the playground and blow soap bubbles as another example of how air fills up space.

Air occupies space.

_____
- - - - - - - - - - - - - - - - - - - -'s Experiment
_____

no air	full of air
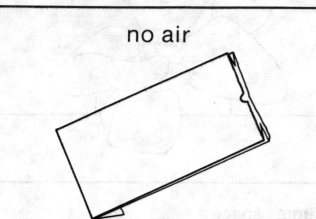	

Concept: Moving air can do work.

Materials:	plastic tubs
	water
	small sailboats

The sailboats can be purchased at a toy store or you and your students can make them.

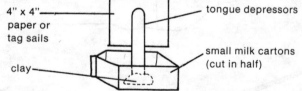

4" x 4"
paper or
tag sails

tongue depressors

small milk cartons
(cut in half)

clay

You will need to cut the cartons in half for your students. They can press the lump of clay into the bottom of the boat, glue the sail to the tongue depressor, and push the end of the tongue depressor into the clay.

A. Discussion:
"How can you get your boat across the water without using your hands?"
You may want to record their suggestions to refer back to after the experiment.

B. Experiment:
1. Put the boat in the water.
2. Blow softly to see if the boat will move. Then blow hard to see how fast and how far it will move.
3. Complete the worksheet. Color the boat and water. Circle the picture at the bottom to show what moved your boat.

C. Share the results:
"Did your boat move by itself?"
"What caused the boat to move?"
"Did it go farther/faster when you blew hard or soft?"
"What are other ways wind can do work?"

Example: flying a kite
moving a windmill
turning a weather vane

Additional Experiments:
1. Race boats to see which is the fastest. Discuss why some boats moved faster.
2. Fill one boat with small objects. See which moves faster, a full boat or an empty one.

Moving air can do work.

_____
- - - - - - - - - - - - - - - - - - - 's Experiment
_____

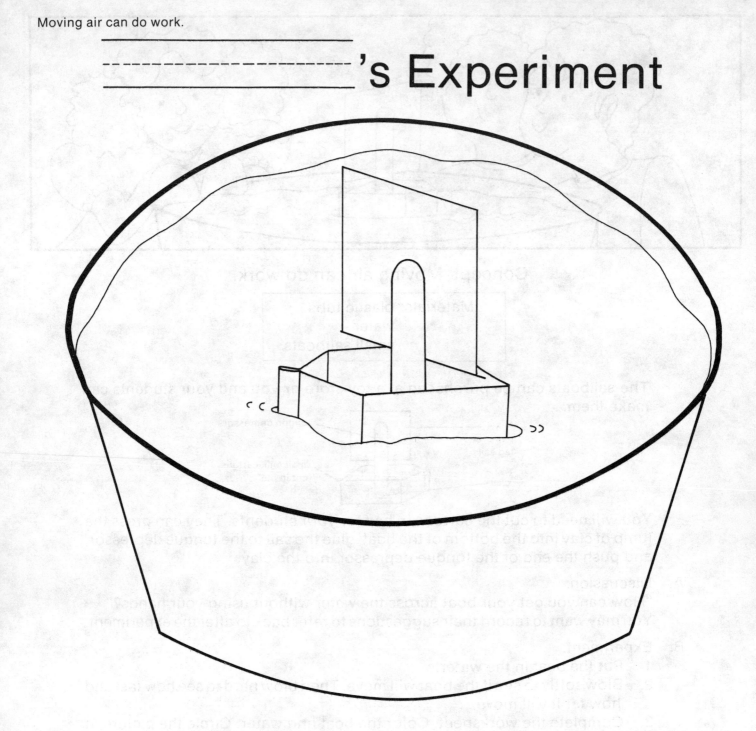

What made our boat move?

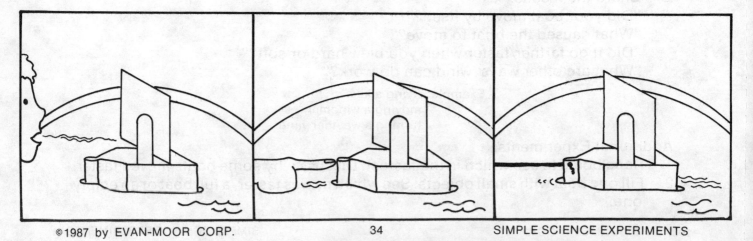

©1987 by EVAN-MOOR CORP.   34

Concepts: Objects move more easily over a smooth surface than a rough one.

Materials: cardboard boxes (filled with objects to give them weight)
              access to smooth surfaces (polished wood floor, linoleum, tile,
              table top, etc.)
              access to rough surfaces (carpet, black top, sand box, etc.)

A.   Discussion:

Discuss the different ways the box might be moved. (Encourage them to include push, pull, and lift.) "We are going to move our boxes over different kinds of surfaces. I want you to think about each one as I name it and tell me if you think it will be easy or hard to push the box across."
Name the ones you have access to:

wood floor	carpet
sand box	playground
table top	tile

You may want to record their predictions to refer back to after the experiment is over.

B.   Experiment:

1.   Have children work in small groups. Push or pull your box over
     _____*(rough surface)*_____ .

2.   Push or pull your box over _____*(smooth surface)*_____ .

C.   Share the results:

"What happened when you moved your box over the rough place?
What happened when you moved your box over the smooth place?
Which was easier ?"

Discuss the worksheet and have children cross out the places that are hard to pull the box over and circle the places that are easy.

©1987 by EVAN-MOOR CORP.   35

Objects move more easily over a smooth surface than a rough one.

_____
- 's Experiment

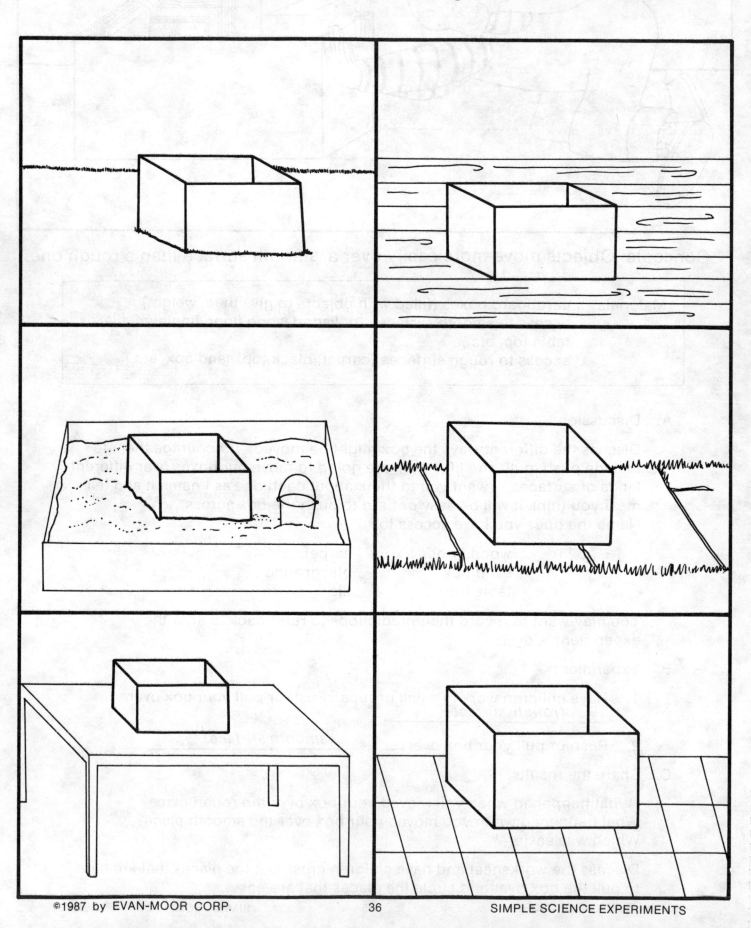

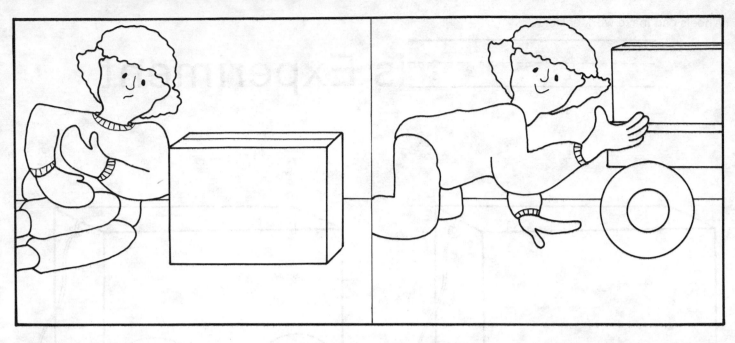

Concept: Wheels can make work easier.

Materials: toy cars with wheels
              toy cars without wheels
              cardboard box - filled with objects to add weight
              wagon

A. Discussion:

Review how having a smooth surface made moving the cardboard box easier. "Can you think of other ways we can make it easier to move our box?" Record any suggestions to refer back to later. "Let's see what happens when we do today's experiment."

B. Experiment:

1. Whole group:
    Push the cardboard box across the room. Pull it back. Now put the box in the wagon. Push the wagon across the room. Pull it back. Discuss which way was easier.

2. Individuals or small groups:
    Try to push the cars without wheels across the floor and carpet. Now try the car with wheels. Which one moves more easily?

3. Record the results of the experiment on the worksheet. Color the objects that moved easily. X the objects that were hard to move.

C. Share the results:

"What happened when you pushed the box by itself? What happened when you put it in the wagon and moved it? Why was this easier? What happened when you tried to push the car with no wheels? Was it easier to move the car with wheels? Can you think of some of the places we use wheels to make our work easier?"

Wheels can make work easier.

_____
----------------------------'s Experiment

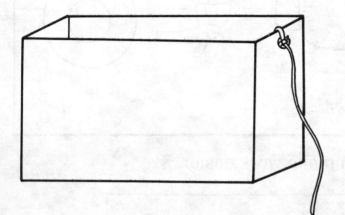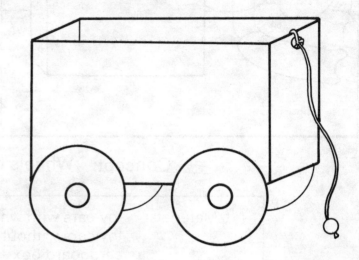

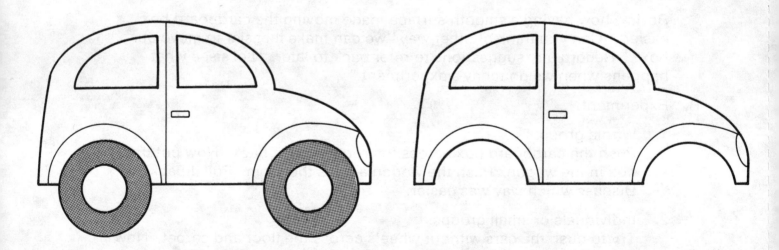

Wheels help.

©1987 by EVAN-MOOR CORP.

SIMPLE SCIENCE EXPERIMENTS

Concept: An inclined plane makes work easier.

Materials: a long board
access to stairs or a table
cardboard box - fill with objects to give it some weight

A. Discussion:

"We need to move this box up the stairs (or up onto the table top). Can you think of a way to make it easy? We learned that wheels can make work easier. Will wheels help us here? I have a long board here. Do you think we can use it to make our job easier? Let's find out."

B. Experiment:

1. Attempt to move the box up the stairs (or table) by pushing, pulling, and lifting.

2. Place the board along the steps (or table) to create an inclined plane. Try to move the box up the stairs now.

C. Share the results:

"Which way was the easier? Can you think of a reason why it was easier to go up the board than up the steps?"

Do the worksheet on page 40.

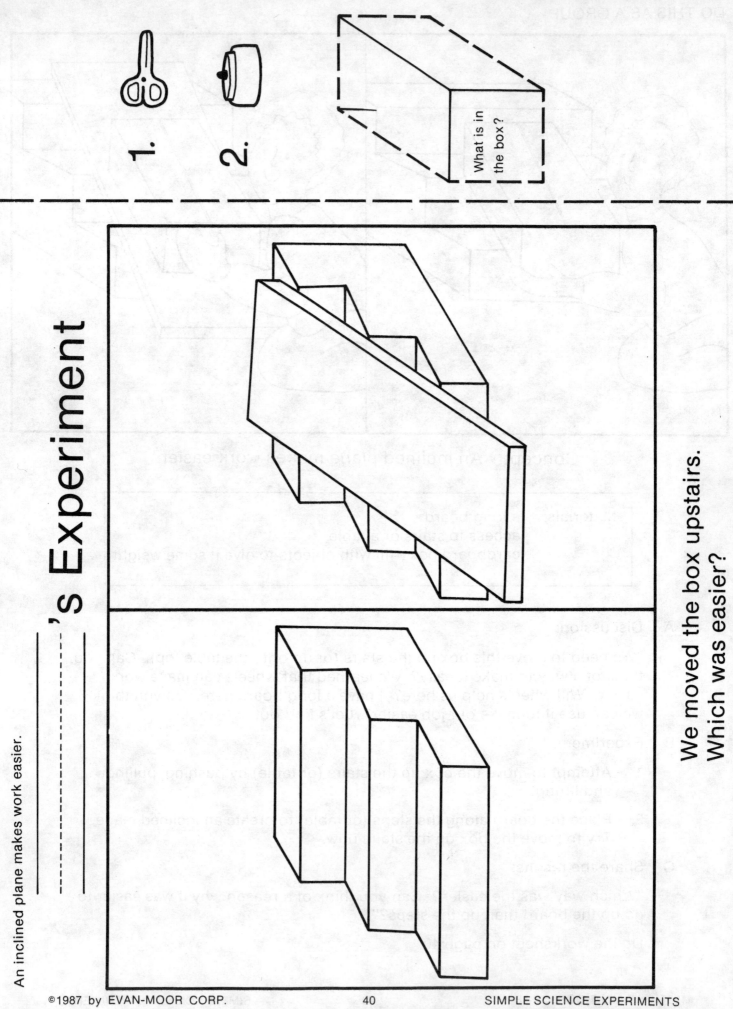

_____'s Experiment

An inclined plane makes work easier.

What is in the box?

1.

2.

We moved the box upstairs.
Which was easier?

Making a Rock Collection
Concept: Rocks can be different sizes.

Materials: egg cartons
             masking tape
             black marking pen
             rocks - assorted colors and sizes

Note: If you are near a rocky area or beach, plan
        a field trip to collect your rocks.

A. Introduction:

    "We are going to be learning about
    rocks. Today you are going to make
    rock collections to use in our
    experiments."

B. Activity:

    1. Each child gets an egg carton.

    2. Put one rock in each hole. (The
        teacher will need to put each
        child's name on masking tape to
        go on his/her egg carton.)

C. Using the collection:

    1. Sort the rocks by size. Do the
        worksheet on page 42. Children
        cut out the rocks and paste them
        from smallest to largest on a
        sheet of construction paper.

    2. Sort the rocks by hardness. Give
        each child a nail. They are to
        scratch each rock with the nail.
        They will not be able to mark the
        hardest rock, but will leave a
        scratch mark on softer rocks.

    3. Sort by color. (See page 43.)

    4. Sort by texture. (See page 45.)

    5. Sort by weight. (See page 47.)

©1987 by EVAN-MOOR CORP.   SIMPLE SCIENCE EXPERIMENTS

Rocks can be different sizes.

_____'s Experiment

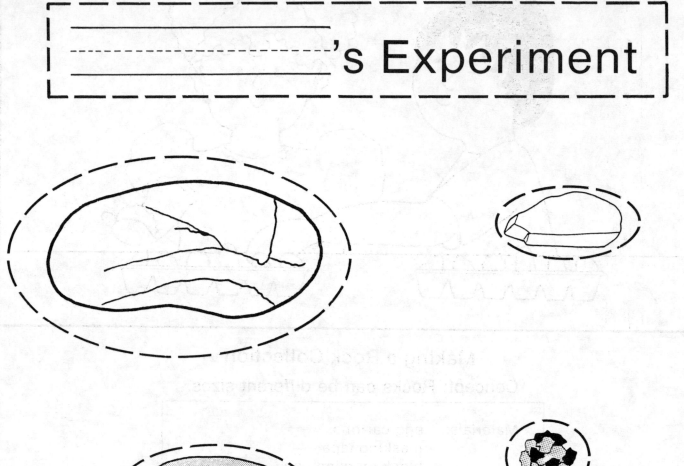

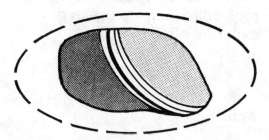

Concept: Rocks are many different colors.

Materials: plastic bowls
             water
             rock collection
             crayons (include gray)
             paper towels

A. Discussion:

"What are some of the way rocks are different? Have you ever found a rock that was a pretty color? How many different colors can rocks be? Let's experiment with our rock collections to find out."

B. Experiment:

1.   Look at your rock collection. What colors do you see?

2.   Put each rock in your dish of water. Did the color change? Can you see the color better? Dry each rock before you put it back in your egg carton.

3.   Make the rocks on your worksheet the same colors as the rocks in your collection.

C. Share the results:

"What happened when you put your rocks in the water? Did anyone have a rock that was _____(color)_____ ? How many different colors did you have? What color are most of your rocks?"

_____'s Experiment

Rocks are many different colors.

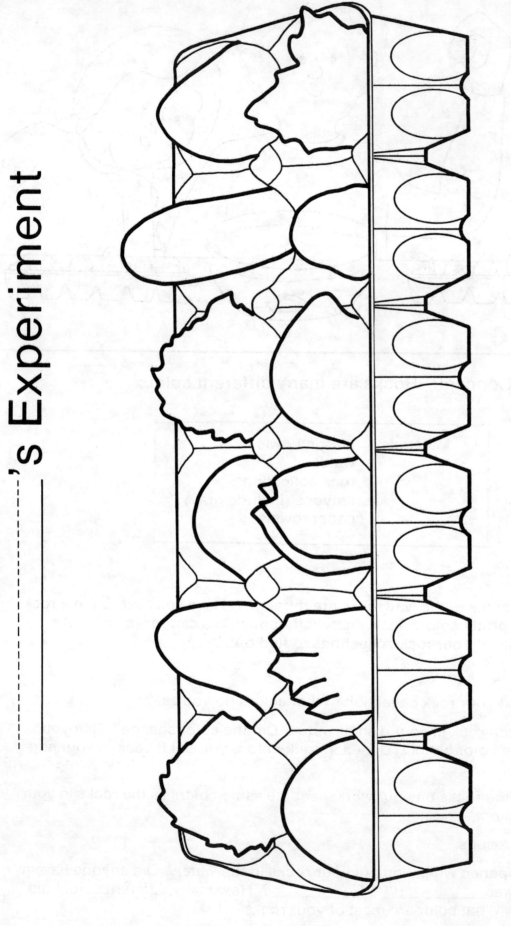

Rocks are different colors.

©1987 by EVAN-MOOR CORP.

44

rough

smooth

Concept: Some rocks are smooth and some rocks are rough.

Materials: rock collection
2 paper plates (or sheets of colored paper)

A. Discussion:

"We know that rocks are different colors. Can you think of other ways rocks are different? Today we are going to look at our rocks and feel them carefully to see which ones are smooth and which ones are rough."

B. Experiment:

1. Look at your collection. Pick up one rock you think will be smooth and one rock you think will be rough. Were you right?

2. Feel each of your rocks. Put all the rough ones on one paper plate and all the smooth ones on your other paper plate.

3. Do the worksheet on page 46. Paste the pictures of smooth rocks together and the pictures of rough rocks together.

C. Share the results:

"Can you think of words that tell how your rocks felt? How many smooth rocks did you have? How many rough rocks did you have?"

Additional Activity:

You may want to discuss the composition of the rocks, looking for rocks that seem to be made up of many types of material and rocks that seem to be only one type of rock.

Some rocks are smooth and some are rough.

_____

---------------------'s Experiment

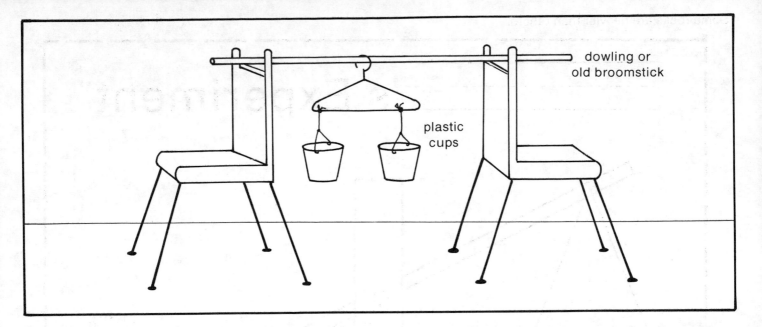

dowling or
old broomstick

plastic
cups

Concept: Some rocks are heavier than others.

Materials: rocks of assorted sizes and weights
              balance scales
              white glue

Note: If you do not have a balance scale in your
       classroom, you can make simple ones
       from coat hangers and plastic cups. They
       will be accurate enough for this activity.

1.   Punch holes in the cups with a hole
      punch.

2.   Attach the cups to the hanger with
      string.

3.   Move the strings back and forth on the
      hanger until the cups are pretty well
      balanced.

A. Discussion:

"We know that rocks are different
colors and that some are smooth and
some are rough. Can you think of
other ways they can be different?
Today we are going to find out which
rocks are the heaviest." Show several
rocks. Have the children try to tell by
first looking and then by feeling,
which rock will be the heaviest and
which one will be the lightest.

B. Experiment:

1.   Have the children work in small
      groups. They need to take two
      rocks at a time and place them in
      the balance scale to see which
      rock is the heaviest. (Let each
      group have several turns.)

2.   Do the worksheet on page 48.
      Children paste the picture to a
      piece of tag or cardboard. They
      then glue a light rock and a heavy
      rock on the cups of the balance
      scale.

C. Share the results:

"Did you find a heavy rock? Was it
the rock you thought would be the
heaviest? Is the biggest rock always
the heaviest?"

Some rocks are heavier than others.

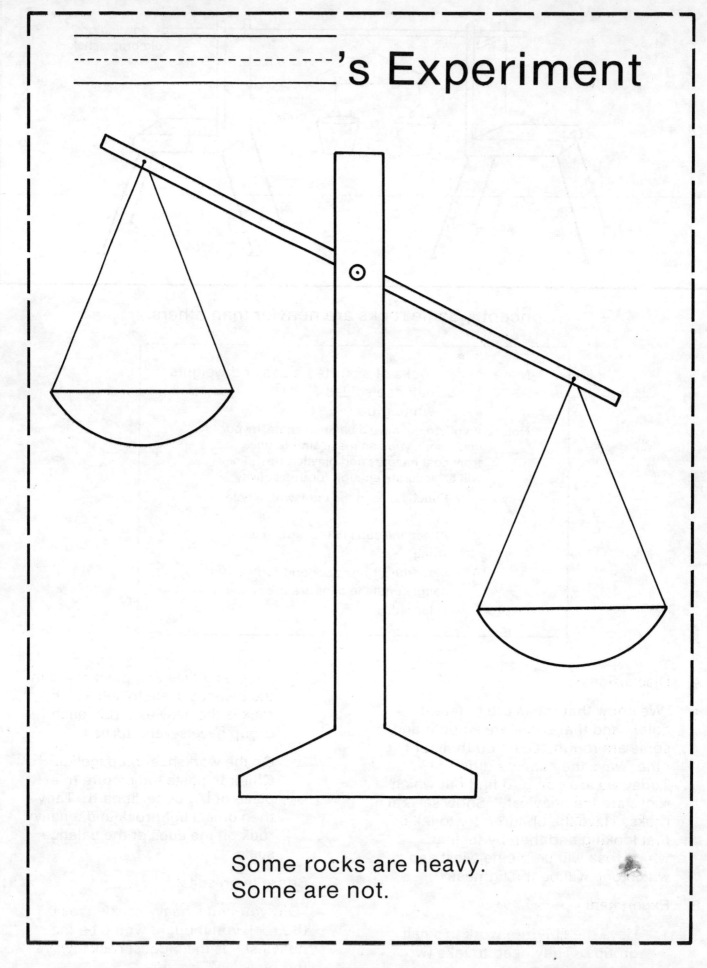

_____'s Experiment

Some rocks are heavy.
Some are not.

©1987 by EVAN-MOOR CORP.

48